THE OP
MATHEI
AN INTI
MA290:

CW01457117

TICS

BLOCK I   MATHEMATICS IN THE ANCIENT WORLD

UNIT 3

THE GREEK CONCEPT OF PROOF

PREPARED BY JOHN FAUVEL FOR THE COURSE TEAM

THE OPEN UNIVERSITY

CONTENTS

This unit forms part of an Open University course. The set book for the course, to which reference is made as **SB**, is:

John Fauvel and Jeremy Gray (editors), *History of Mathematics: A Reader*, Macmillan 1987.

The Open University, Walton Hall, Milton Keynes.

First published 1987. Reprinted 1989

Designed by the Graphic Design Group of the Open University.

Typeset in Great Britain by Santype International Ltd, Salisbury.

Printed in Great Britain by BPCC Wheatons Ltd, Exeter

ISBN 0 335 14247 8

This text forms part of the correspondence element of an Open University Second Level Course.

For general availability of supporting material referred to in this text, please write to Open University Educational Enterprises Limited, 12 Cofferidge Close, Stony Stratford, Milton Keynes MK11 1BY, Great Britain.

Further information on Open University courses may be obtained from The Admissions Office, The Open University, P.O. Box 48, Milton Keynes MK7 6AB.

1.2

3.1 THE DEVELOPMENT OF PROOF

If there is one aspect of Greek mathematics which more than any other distinguishes it from the mathematical activity of earlier cultures, it is the notion of *proving* results. In the last unit you saw several examples of proofs from different stages of Greek mathematical development. This aspect of mathematics did not, of course, spring up overnight, nor was it divorced from other social, political and ideological developments in classical Greece. In this section we discover what we can of its development—within the limits, as ever, of what sources remain—in what seems to have been the critical period of the sixth to the fourth centuries BC, from Thales to Euclid. That period ends with the imposing and influential axiomatic, deductive, structure of proved results that make up the thirteen books of Euclid's *Elements*. It starts somewhat mistily, with what are probably legendary attributions to Thales; let us look at one of these now, as it will lead us usefully into the subject.

Proclus, writing in the fifth century AD, reported

> The famous Thales is said to have been the first to demonstrate that the circle is bisected by the diameter. The cause of this bisection is the undeviating course of the straight line through the centre.

This is an intriguing claim. Proclus probably took it from the now-lost *History of Geometry* by Aristotle's pupil Eudemus, written 200 years after Thales (and 800 years before Proclus), and it seems best to take it as a fourth century BC perception of what might have, or should have, happened back in the time of Thales. But it is intriguing because it encapsulates so nicely the unprecedented Greek approach to mathematics. At an earlier stage it may well not have occurred to anyone that such a statement—that the diameter of a circle bisects the circle—would need proving; nor indeed would they necessarily have the concept of proof in order to formulate that thought. But even once the idea of proving things exists, it is far from obvious what one is to do next, or what a proof should look like.

Question 1   How would *you* prove that a diameter bisects the circle, or to put it another way, if you believe it, *why* do you believe it? Jot down the thoughts which come into your mind when faced with this problem.

Comment _____

It may be that the property seems so 'obvious' that the idea of proving it is rather baffling. It is hard to see what other property to start from, in order to *deduce* the one we are interested in. If it can be proved true for one particular circle, is it thereby shown to be true for all circles, whatever their sizes?

And there are other problems: is it clear just what the property is? What is meant by bisecting the circle? Dividing it into equal areas, or equal shapes, or two bits of circumference that are equal in length? These amount to the same thing in this instance, but how these possibilities relate in general forms a fresh set of problems, opened up by trying to decide even what the *definition* of our property is.

One thing we could do is to draw a circle on paper, cut it out and fold it over along a diameter. Then the two halves coinciding would show (depending, perhaps, on our scissors and folding skills) all three of the possible properties held: equal areas, equal shapes and equal edge lengths. But then this procedure raises another question, of the relation between physical bits of paper and 'the circle', which sounds a more abstract concept. This experiment may lead one to *believe* the result, while leaving doubts that one has *proved* it.   ■

Attempting this exercise may have persuaded you that the notion of proving things involves quite a complicated set of considerations. Unravelling and clarifying these historically will occupy us for the remainder of this section. Before leaving this property of the circle, observe an ironic twist to any attempt to unravel the history of the problem. If we turn to Euclid's *Elements* to discover how he proved it, we find that he side-stepped our problem by incorporating it as part of the *definition* of a circle's diameter. So—whatever Thales may or may not have done—Euclid did not feel he needed to prove it at all (though the result can, in fact, be proved, as a consequence of propositions proved later in *Elements* III).

See *Elements* I, Def. 17, i.e. Book I, *Definition* 17, in **SB** 3.B1.

3

One reason for scepticism that Thales, back in the sixth century BC, 'demonstrated' any mathematical results is that there is no reliable supporting evidence. Furthermore, such a claim does not fit in with our understanding of general philosophical developments in the *subsequent* century. The earliest example of sustained deductive argument we know of occurs in the poem by Parmenides (early fifth century BC) called *The Way of Truth*. Parmenides lived in Elea, in southern Italy, and we infer when he lived because Plato relates that Parmenides, when quite old, once visited Athens with his pupil Zeno, where he met the young Socrates; and this would have been about 450 BC. There is an extract from Parmenides' poem in the Source Book (**SB** 2.C1); Please *glance through that briefly now* to see the kind of style and reasoning involved. Plato, *Parmenides* 127A

Now, although not the most pellucid text you will meet, you could probably see that it is about how we know things, and appeals explicitly and self-consciously to deductive argumentation and logical principles. What Parmenides established or consolidated is a fundamental distinction between two sorts of knowledge: that acquired through the senses ('the opinions of mortals, in which there is no true reliance') and that attained through reason alone ('the unshaken heart of well-rounded truth'). The latter is the only knowledge that can be trusted, the only *true* knowledge (so the other sort would be better described as 'belief' or 'opinion'). This is what is taken to lie behind Parmenides saying, for instance: 'judge by reason (*logos*) the strife-encompassed refutation spoken by me'. Knowledge, as opposed to opinion, has a 'force' or 'compulsion' about it—Parmenides speaks of 'the force of conviction'. That true knowledge is something to be contrasted with beliefs acquired from the everyday world, is a startling claim which has underpinned and haunted much Western thought ever since.

During the fifth and fourth centuries, other philosophers discussed and clarified the distinction and its consequences for modes of argument, as well as using it in a taken-for-granted way. Thus the *Meno* passage you read at the very beginning of the last unit—written within a century of Parmenides' poem—had quite a lengthy discussion about 'opinions' and 'knowledge', how to get from one to the other and what their status is. Indeed, Plato put into Socrates' mouth logical structures precisely akin to the use Parmenides made of the phase 'it is or it is not':

> SOCRATES   If then there are going to exist in him, both while he is and while he is not a man, true opinions which can be aroused by questioning and turned into knowledge, may we say that his soul has been for ever in a state of knowledge? *Clearly he always either is or is not a man.*   **SB** 2.E1

So it is in the fifth century BC that startling and profound developments took place in two related aspects of thought: in the status of different kinds of knowledge, and in the techniques for acquiring—or persuading others of—true knowledge. The latter concerns the question of proof, the former what the proof relates to or achieves.

Where, then, does mathematics fit into this picture? We should bear in mind that what we now identify as the mathematics of the period was not a separate field of study to the extent that it is today, but was part of the general realm of intellectual discourse which had political, moral, cosmological and also other aspects to it. There were, though, schools or traditions of teachers and learners which might emphasise one or other aspect or have particular preoccupations.

Both the *Eleatics* and the *Pythagoreans* flourished in southern Italy in the fifth century. The *Eleatics*—Parmenides of Elea, and his followers such as Zeno—seem to have emphasised logical concerns in particular. The *Pythagoreans* developed thought of a more religious and ethical cast, with mathematical overtones whose precise nature is still disputed. And in Athens during this same century were *Sophists*, teachers of argument and rhetoric, against whose style and approach Socrates and Plato argued eloquently. There seems to have been a wide measure of agreement, though, that mathematical knowledge was special and of a higher kind of certainty than the everyday knowledge received through the senses. Mathematical knowledge was therefore seen to be inextricably connected with the way it was acquired.

In order to study the development of styles of mathematical proof, as opposed to the development of the idea that proving mathematical results is a desirable and necessary aspect of doing mathematics, we need to undertake some historical reconstruction and inference. For the earliest fully recorded proof, Hippocrates' quadrature of the lune, is in quite a sophisticated and logically complicated style, as you saw in *Unit 2*. So it is reasonable to infer that there were simpler methods of justification earlier in the fifth century. (We shall not be too precise about the dating, partly because we do not know and partly because it may be that Eudemus' account of Hippocrates' work introduced later ideas.) In fact, the discussion between Socrates and the slave boy, in *Meno*, culminates in what seems to be an earlier proof style. Recall that Socrates drew the final diagram in the sand, from which the boy could essentially *see* (with the aid of a few leading questions) what the answer was and why it must be so.

Note that there are two aspects to this: finding out some mathematical result, and demonstrating that the result is true. We are mostly concerned with the latter aspect at the moment, but will return to these two aspects later.

Question 2 Study that passage again (**SB** 2.E1, from 'Socrates here rubs out the previous figures . . .' onwards) in order to ascertain how rigorous you consider the proof to be. Are there any steps that you think should have been proved more carefully?

Comment ———————————————————————————

You may have noticed that Socrates gains the boy's ready assent to the assertion that the diagonal of the square bisects it ('Now does this line going from corner to corner cut each of these squares in half?'), a proposition analogous to that which Thales is said to have proved for the circle. It is, indeed, fairly 'obvious' that this is true. History does not record how Socrates would have responded had the boy replied 'No—prove it to me!' ■

The style of justification in which things are presented so that one can immediately 'see' that the result is true are called *diknume* proofs. We saw two geometrical examples in the *Meno* passage, the overall perception of the final diagram and the case of the diagonal bisecting the square (though the latter slipped past so quickly it may be stretching a point to claim it as proved at all). Some early results about numbers were probably first proved in *diknume* style, for instance knowledge about odd and even numbers. (According to Aristotle, investigation of 'the odd and the even' was of particularly fundamental significance to the Pythagoreans.) Let us consider the claim that the sum of any two odd numbers is an even number; it can easily be checked for individual cases ($3 + 5 = 8$, $11 + 7 = 18$, and so on) but that does not amount to proving the general statement. The *diknume* proof arises in considering the representation of numbers as, say, pebbles. Then if two odd numbers are added—if the pebble patterns are joined up—one can *see* that the resultant number must be even.

The word *diknume*, pronounced with three syllables to rhyme with 'roomy', comes from the Greek meaning 'to show something to be so, to make visible or evident'.

There is evidence in a fragment of a work by Epicharmus, mid-fifth century BC, that numbers were represented as pebbles and in such a way that the oddness or evenness was apparent.

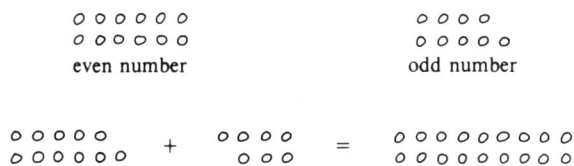

Figure 1

Evidence for the antiquity of these concerns and perhaps a proof such as this can be inferred from examining the treatment of the *odd* and the *even* in Euclid's *Elements*. Examine *Elements* IX, 20–22, in your Source Book (**SB** 3.D2), and answer this question.

Elements IX, 20–22 means Book IX, *Propositions* 20–22.

Question 3 Generally Euclid orders his propositions so that the logical structure is built up carefully. (We look at this later in the unit.) Do you feel that to be a fair description of these three consecutive propositions? How closely does any of them come to a *diknume* proof?

Comment

You studied Proposition 20 in the last unit (Question 10), so you may remember it had quite a sophisticated logical structure. Both the proof, and the proposition being proved, are patently more advanced and conceptually difficult than the results about even and odd numbers in Propositions 21–22. The latter appear to be tagged on as a kind of afterthought.

Despite the fact that Euclid is representing numbers as lines, not as dots or pebbles, it is not difficult to see the proofs of Propositions 21 and 22 as translations into a more overtly logical framework of arguments in the *diknume* style. If the proof of Proposition 22 arose in this way, something a bit more elaborate was done with the pebbles than in the case of just two odd numbers that we surmised above, but the translation is still fairly clear. ■

So partly from their positioning which suggests old material he had no particular use for in his logical framing of number theory, and partly from the simplicity and directness of the proofs, it could be argued that Euclid incorporated into the *Elements* some fifth century results on 'the odd and the even', with something plausibly close to their original demonstration.

There are two major problems with *diknume* proofs, though. However convincing a proof may be in itself, each new result has to be considered afresh on its merits. Because of the directness and immediacy of each proof, it is difficult to build up in this way a structure in which results depend on other results. Secondly, they appear to depend on sense impressions—on literally *seeing*—in a way that lays itself open to philosophical critique that the proof-style is too closely tied to the transient everyday world of human opinions, rather than the world of mathematical truth. This need not be a fatal objection—it can be argued that the pebbles or the sand-diagram are but aids to enable inner reason to attain true knowledge—but the style is clearly different from the path to knowledge advocated by Parmenides and the Eleatics.

Is it possible to prove things with absolute certainty, with pure reason alone, making no concessions to knowledge aroused by the senses? The Eleatics appear to have thought so, and the proof style introduced by them, *indirect proof* or *reductio ad absurdum*, turned out to be enormously influential in Greek and later mathematics. You saw three examples of this approach in the last unit: in the Euclidean proof that the number of primes is unlimited (*Elements* IX, 20); in the proof referred to by Aristotle that the side and diagonal of a square are incommensurable (*Unit 2*, Section 3); and, in an informal way, as part of the discussion between Socrates and the slave boy in *Meno* (point (v) of *Unit 2*, Section 1).

The historical question of whether mathematicians adopted this proof style as a direct consequence of the philosophical arguments of Parmenides and his followers is unresolved. The best we can say is that indirect proofs were employed in mathematical argument some time after their philosophical advocacy by the Eleatics. What they rely on are purely logical assertions whose truth no-one could deny, regardless of the state of the world or one's sense impressions. You saw that Parmenides argued from the statement 'it is or it is not', and that later, in *Meno*, Socrates uses 'Clearly he always either is or is not a man'. Similarly the mathematical examples you saw in the last unit arise from, respectively,

> Either G is a prime other than A, B, C, or it is not. (Euclid)
> Either the diagonal and the side are commensurable, or they are incommensurable. (Aristotle)
> To double the square, either you double the sides, or you do not. (Plato)

All of these statements must receive universal acceptance, regardless of which alternative is actually the case. They are truths of pure reason, and thus a suitable starting point for attaining true knowledge by deducing further truths by the use of reason alone. The way this deduction works, in *reductio* proofs, is to take one alternative, and disprove it; from this it follows that the other alternative must be true. Disproving the alternative is again a matter of pure logic; you deduce something which contradicts what you already know or have laid down, which is therefore a logical absurdity. The first *reductio* proof in Euclid's *Elements* is a good example.

Question 4 Study Euclid's *Elements* I, 6 in your Source Book (**SB 3.B3**), and answer the following questions.

(i) What is he trying to prove?

(ii) What is the equivalent here of the '*Either* something *or* not that something' statement?

(iii) So which alternative will he take in order to disprove it?

(iv) What is the absurdity or contradiction he arrives at through pursuing this alternative?

Comment ───────────────────────────────────

(i) That in a triangle with two equal angles, the sides opposite those angles are themselves equal.

(ii) *Either* the two sides in question are equal, *or* they are unequal.

(iii) Euclid takes the 'they are unequal' alternative and deduces a contradiction from that.

(iv) He shows that on this assumption two different triangles (*ABC* and *DBC* in his diagram) are equal; but it is clear from the existence of the triangular sliver *ACD* that *DBC* is actually smaller, by that amount, than *ABC*. That one triangle is, simultaneously, equal to and smaller than another is a logical absurdity. ■

There is an important point to notice about *reductio* proofs. The initial statement which generates the proof structure, of the 'it is or it is not' kind, has to consist of incompatible alternatives which exhaust all the possibilities. Otherwise, disproving one alternative would say nothing of consequence about the other. For instance, in the case above it would be unavailing to set out from the statement 'either the two sides are equal or the two angles are equal', as these are not incompatible. To set out from 'either *AB* equals *AC* or *AB* is less than *AC*' would be equally fruitless, since it does not take account of the possibility that *AB* is greater than *AC*.

In analysing the nature of *reductio* proofs we are running ahead of the history somewhat. We cannot be at all sure that mathematicians in the fifth century thought *about* the nature of proof in the explicit way we have just done, an analysis which owes more to the work of Aristotle in the middle of the following century. It is not until Aristotle that we find a full and careful analysis of deductive argument and modes of proof, so mathematicians earlier may have operated on a more intuitive and less self-conscious plane, for all we know to the contrary. A similar difficulty arises in considering what fifth-century conceptions of mathematical knowledge may have been—here the later influence of Plato is so strong as to colour inescapably our perceptions of his predecessors.

3.2 PROOFS BY CONSTRUCTION

Although *reductio* proofs are very powerful, and were and are much used, they have an undesirable consequence when used to attain mathematical knowledge, compared for example with the state of mind achieved through a *diknume* proof. With the latter you can see clearly and convincingly *why* the result is true, whereas *reductio* proofs demonstrate unarguably *that* the result is true, but do not necessarily bring understanding of why or how. What has been proved is that things cannot be other than they are, that other possibilities are absurd, which is not the same as having direct knowledge of why a result is, in fact *has* to be, the case. The Greeks did indeed develop other proof styles, more logically convincing than *diknume*, leading to deeper understanding than did the indirect *reductio* style. Logical proofs involving longer chains of direct deduction were in use by the latter part of the fifth century, as you saw in Hippocrates' quadrature of the lune (*Unit 2*, Box 1). In that instance, the proof involved a *construction* of the lune whose quadrature was sought. But even when the object you were seeking to prove something about was already given, proofs often involved further constructions on which a deductive argument would be based. Let us look at an example to see what is involved. Please *read Euclid's Elements* I, 32 (**SB** 3.B4) and the accompanying extract from Proclus.

These are both proofs, using different constructions, that the sum of the internal angles of a triangle equals two right angles. In the Euclidean proof, a side is extended in one direction, and at that corner a line drawn parallel to the opposite side. The other construction, ascribed by Eudemus to 'the Pythagoreans', consists just of drawing a line through one corner of the triangle, parallel to the opposite side. That done, in both cases the rest of the proof consists of direct reasoning about the figure thus constructed.

The third extract in **SB** 3.B4 is a brief reference by Aristotle to a proof of this proposition (as usual, in illustration of a more general point). The remark about the parallel being 'drawn up' suggests it was the proof appearing (later) in Euclid which he had in mind, though otherwise he could be referring to either. What Aristotle says is interesting in suggesting that after the construction has been done, the result follows by the process we have described as *diknume*. Do you agree?

Question 5 Reread the two proofs, compare them with what Aristotle says and formulate your thoughts on what is to be said for and against Aristotle's analysis.

Comment ───

What Aristotle's discussion brings to light is something we rather skated over in our earlier account of *diknume* proofs; *seeing* something to be obviously true depends on the state of mind and knowledge you bring to bear on the perception. You saw in the *Meno* proof, too, that Socrates needed to ask several questions to awaken in the slave boy his own 'personal opinion' that the result is true.

In the case of the internal angle-sum of the triangle, the result, and the reason for it, are certainly 'clear from merely looking at the figure', *provided* that one is already in a state of knowledge in which several other results are so well known as not to need thinking about: knowledge of parallel lines, of the relation between particular angles, knowledge of adding angles just by looking at them, and so on. In this sense, Aristotle's claim is right.

On the other hand, the detailed argument given by Euclid or by Proclus is necessary to establish a logically convincing proof, which can be shared with others more readily than can an inner conviction. The proofs spell out in more detail what it is about the constructed situation that leads to the *diknume* knowledge that the result is true. Without the detailed logical argument there is no assurance that one's feeling of clear and certain knowledge is not mistaken. ■

So proofs involving constructions can be described as a sophisticated version of *diknume* proofs, in which the initially given mathematical object is changed or added to by construction, into one enabling the truth of the proposition to be seen more readily, or argued towards more easily. We have not yet specified very clearly what 'a construction' is, however.

The constructions in the above proofs involved extending lines further, and drawing lines parallel to other lines. The construction in *Meno* consisted of drawing three further squares and drawing a line diagonally across each. In Hippocrates' quadrature he constructed the lune by drawing arcs of circles in a particular relationship to a certain triangle. What these constructions have in common is that they can all be effected by just two instruments, an unmarked ruler—a straight-edge—and a pair of compasses, a device for drawing circles.

We shall generally refer to the unmarked ruler as a 'straight-edge'.

The idea of constructions exemplified in this way has a somewhat practical-sounding air. We seem to be in the realm of everyday physical activity, far from the disembodied world of pure reason that was beginning to be taken as the mathematical ideal. Yet we can abstract from the physical instruments and actions, and consider their effects—namely, straight lines and circles—as pure geometric objects. It is not known how fully this point was debated among mathematicians in the fifth century, but it does seem to have been in the latter part of that century that the idea emerged of seeing what could be constructed by using only a straight-edge and pair of compasses. This idea proved very fertile, both in the remarkable number of things that can be done using these apparently restricted means, and—as we shall see later—for stimulating investigation of problems that proved recalcitrant by these means. Indeed, the emergence of straight-edge and compasses (or line-and-circle) constructions signalled the beginning of the much deeper interest in geometry that characterised the following two centuries.

Let us look at an example of a line-and-circle construction in more detail, to see what is involved. In the eventual construction of the *Meno* diagram, Socrates added three more squares to the starting one; we can imagine him to be making rapid free-hand strokes in the sand, gaining the boy's concurrence at each stage:

SOCRATES   Now we can add another equal to it like this?
BOY   Yes

(and so on). But what is involved in the logical justification for this claim? Turn to *Elements* I, 46 in the Source Book (**SB** 3.B6), where Euclid describes (at what you may feel is surprising length) the construction of a square.

This proposition falls into two parts: the first few lines (down to 'and through the point B let BE be drawn parallel to AD') describe the actual construction of the figure, then the rest is the demonstration that the figure thus constructed *is* a square.

Box 1   The construction of a square (*Elements* I, 46)

1   From the starting line, labelled *AB*, draw at *A* a line at right angles.

Proposition 11 shows how this can be done.

2   Mark off point *D* on the new line, at the same distance from *A* as *B* is.

By setting the compasses, with the point at A, to the length AB and swinging it round to the new line.

3   Draw from *D* a line parallel to *AB*.

Proposition 31 shows how this can be done.

4   Draw from *B* a line parallel to *AD*.

Ditto.

5   Label the point where the two latest lines intersect as *E*.

Then *ADEB* is the figure constructed, and it remains to show that it is a square.

Euclid proves successively that it is a parallelogram, so opposite sides are equal (by Proposition 34); and that in this case all four sides are equal; and that all four angles are right angles. These last two conclusions form the definition of a square.

Follow this through either in your mind's eye, or in practice if you have a set of geometrical instruments left over from your school-days: you need just a ruler, a pair of compasses and a pencil.

It should be noticed how direct argument about a constructed object is intuitively more appealing than an indirect *reductio* proof: one emerges after studying it with a clear perception of the square as an object which can be constructed, which is more satisfying than the assurance provided by a *reductio* proof that things could not be otherwise. But notice too that it did not start from an undeniable logical truth; some things have to be taken for granted. Here lines and circles are the building blocks from which everything is constructed.

This is a very interesting departure. It seems that some fifth century Greek mathematicians took the straight line and the circle to be fundamental geometric objects. The line is the elementary thing out of which all other rectilinear geometric objects are made (the square, the triangle, and so on), and the circle is the elementary curved thing, with the property that all its points are the same distance from the centre. From these two building blocks the mathematician can then construct further objects and describe properties of those objects, deducing new results from previous ones. This ensured knowledge that was both logically secure and also grounded in objects convincing to the senses.

Line-and-circle constructions were those most frequently used; in Euclid's *Elements* almost no others are found. But other kinds of construction were considered too, using instruments other than compasses and an unmarked ruler. Most notably, there was a construction called *neusis*, the effect of which is to insert a line of pre-determined length between two other (straight or curved) lines so that it points in a given direction, towards a fixed point. In physical terms this can be thought of as moving around a marked ruler (one on which the desired length has been marked) until it passes through the fixed point and the marks coincide with the two lines. The earliest indication of this construction is in Hippocrates' quadrature of lunes. Please *turn now to* **SB 2.B1** (Hippocrates' Quadrature of Lunes) and read paragraph 4. Hippocrates is now dealing with another lune, this time one whose outer curved edge is less than a semicircle.

Neusis is sometimes translated as verging.

The *neusis* occurs in the 'preliminary construction', prior to constructing the lune itself. The diagram is less terrifying than it looks if we take it step by step. (We go through the construction in detail, partly to learn about it and partly to help you gain confidence in working through geometric proofs.)

> ... Let there be a circle with diameter *AB* and centre *K*. . . .

This can be drawn using a pair of compasses. (Actually, a semicircle is sufficient for the construction, and is all that appears in Figure 2.)

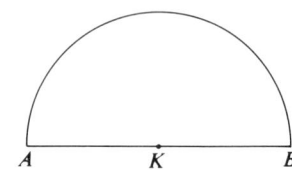
Figure 2

> ... Let *CD* bisect *BK* at right angles; . . .

A straightforward straight-edge and compasses construction. (Figure 3)

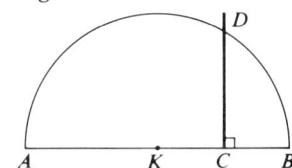
Figure 3

> ... and let the straight line *EF* be placed between this and the circumference verging towards *B* so that the square on it is one-and-a-half times the square on one of the radii. . . .

This is the *neusis*. Imagine a ruler with the distance *EF* marked on it, where $EF^2 = 1\frac{1}{2}AK^2$, which is then moved about, always so as to pass through *B*, until *E* lies on the circumference and *F* on the vertical line, as in Figure 4.

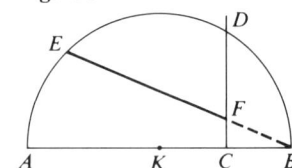
Figure 4

> ... Let *EG* be drawn parallel to *AB*, . . .

A straight-edge and compasses construction, as we saw above (*Elements* I, 31). (Figure 5)

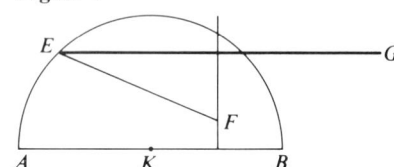
Figure 5

> ... and from *K* let straight lines be drawn joining *E* and *F*. . . .

A straight-edge construction! (Figure 6)

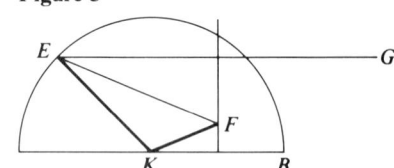
Figure 6

> ... Let the straight line *KF* joined to *F* and produced meet *EG* at *G*, . . .

'Produced' means the line is to be extended, and wherever it cuts the parallel is to be labelled *G*. (Figure 7)

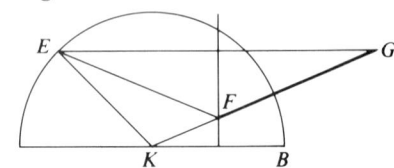
Figure 7

> ... and again let straight lines be drawn from *B* joining *F* and *G*. . . .

This concludes the 'preliminary construction'. (Figure 8)

> ... It is then manifest that *EF* produced will pass through *B*—for by hypothesis *EF* verges towards *B*— . . .

This is pointing out that the line which has just been drawn from *B* to *F* is in the same straight line as *EF*, because of the *neusis*.

> ... and *BG* will be equal to *EK*.

This is not at all 'manifest' in the way that the preceding claim is. It *is* true, but its justification depends on propositions about the geometry of triangles (such as are found later in Euclid's *Elements* I). So this claim illustrates what we learnt in the discussion of Aristotle above (Question 5)—the result that the lengths *EK* and *BG* are equal is clearly true, in a *diknume* way, to someone who already has quite a bit of geometrical knowledge. Such a person could see immediately that this was true, where someone with less experience would need to work more consciously through a train of reasoning.

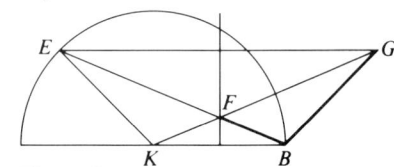
Figure 8

We summarise the rest of the proof more succinctly. After this 'preliminary construction', Hippocrates can straightaway construct the lune, as the space lying between two circular arcs. The outer arc goes through *EKBG*, the inner through *EFG*. Then he argued that (the area of) this lune (Figure 9) is equal to (that of) the rectilinear figure *EKBGF* (Figure 10), thus effecting the quadrature. The argument is similar to that in the previous quadrature complicated by the fact that there are now *three* little outer arcs, but only *two* little inner ones. But here the *neusis* condition comes to the rescue, for that ensured that each little inner arc is one-and-a-half times each outer one (Figure 11).

You have seen, then, that a quite complicated diagram, involving both line-and-circle and *neusis* constructions, was used in the work of Hippocrates, about 450–430 BC. You have also seen that the purpose of a construction proof is to build up a geometrical situation on which an argument can be based to show the truth of the result required. In the course of the argument, other (previously proved) geometric knowledge will be brought in, both to make deductions about the geometric situation and to ensure that the constructions are valid and have the required properties.

The status of *neusis* constructions is a bit puzzling. On the one hand they seem rather less fundamental than the line and circle constructions, and when physically realised in juggling marked rulers about are perhaps intuitively unappealing. On the other hand, some proofs are possible by using *neusis* construction which cannot be done by line-and-circle, though the example we have been looking at from Hippocrates is not such a case, as it happens. (It is possible to devise a straight-edge and compasses construction to draw a line of the required length in the right place, using techniques later set down in Euclid's *Elements* II, for instance.) It is notable, however, that neither Hippocrates nor his commentators Eudemus or Simplicius felt it necessary or desirable to 'reduce' the *neusis* construction in this way. Also there is a passage in Aristotle implying that *neusis* construction was accepted as a fundamental part of geometry. Talking about how the first principles of mathematics must be assumed, Aristotle went on to claim:

> thus arithmetic assumes the answer to the question what is [meant by] 'odd' or 'even', 'a square' or 'a cube', and geometry to the question what is [meant by] 'the irrational' or 'deflection' or '*verging*' (*neuein*); but that there are such things is proved by means of the common principles and of what has already been demonstrated.

In the next section we shall see another example of a *neusis* construction, in connection with the Greek investigations of problems which apparently could not be solved by line-and-circle constructions.

We conclude this section by looking at a distinction raised briefly earlier, the difference between discovering and proving mathematical knowledge. It will probably have struck you when working through Hippocrates' construction just now that the condition on the *neusis* (namely, that the square on one line was one and a half times the square on some other line) was strangely unmotivated, until the point of it became evident at the end. Clearly Hippocrates did not just kneel in the sand one day and start sketching out the construction as we have it. He must have been working backwards, in some sense, in order to know that it was precisely this condition he would want for the logical argument later.

This 'working backwards' was formalised by Greek mathematicians, under the name *analysis*. The fullest account we have of this was given many centuries later, by Pappus (c. AD 320): please *read his description now*, **SB** 5.B3, paragraphs 2 and 3.

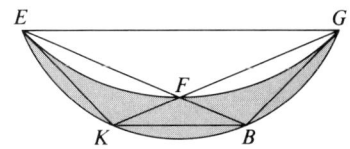

Figure 9

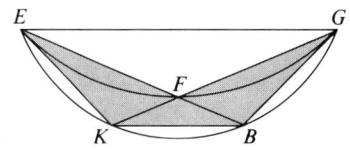

Figure 10

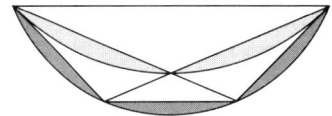

Figure 11

SB 2.H1(a)

Question 6 In that passage Pappus describes the difference between the methods of *analysis* and *synthesis*. Which of these would you say is the approach followed in the quadrature of Hippocrates?

Comment ───────────────────────────────

The distinction as Pappus puts it is a bit hard to follow because of the rather technical, logical, language he uses. In *synthesis* 'we finally arrive at the construction of what was sought', whereas in *analysis* we take 'that which is sought as though it were admitted': so the proof you have been studying is synthetic, working *from* generally agreed starting points ('Let there be a circle . . .' etc.) *towards* the construction and proof of what is required.

We may suppose that this stage was preceded by one amounting to 'analysis', in which he analysed the conclusion he wanted to reach, breaking it down in stages until he got to something known to be true or allowable, such as *neusis* or line and circle constructions. Only by some such process, however informal, could Hippocrates have known in what direction to take the subsequent synthetic proof. ■

Almost all the Greek proofs we have are synthetic. If the analytic method were indeed regularly used in conjunction with the synthetic one then most traces of it have been lost. Indeed, the synthetic proofs that remain (particularly those by Archimedes and Apollonius that we look at in the next unit) presented such an unmotivated appearance to the sixteenth- and seventeenth-century mathematicians who studied them as to suggest that 'the ancients' had a hidden method of discovery. For otherwise, it was not clear how they could have known their synthetic proofs were going in the right direction.

We shall come back to *analysis* in subsequent blocks; later uses of the word by mathematicians do not carry precisely its Greek connotations, but are not entirely unrelated either.

3.3 THE THREE CLASSICAL PROBLEMS

We have seen that towards the end of the fifth century BC construction techniques were increasingly used for reaching and justifying mathematical results. This went hand-in-hand with an increasing concentration on *geometrical* aspects of mathematics. Recall that in *Unit 2*, Section 2, we spoke of a research tradition as comprising *problems*, *methods* of approach, and means of *justification*, and what you are now seeing is how interrelated these things are. The mathematical problems tackled, and indeed the whole geometrical bias of so much Greek mathematics, both stimulated and were stimulated by construction methods as a mathematical style. Thus at quite an early stage the problem arose of what geometrical problems could not be solved through the fundamental constructions. Soon the realisation seems to have dawned that some seemingly straightforward problems were not amenable to solution by means of the basic straight-edge and compasses constructions. There were three such problems in particular, the so-called *three classical problems*.

1 Trisecting the angle.
Given any angle (between straight lines), to construct two lines which divide the angle into three equal parts. (This problem reduces to finding *one* of these lines, for if that can be constructed then the other line would arise immediately by bisecting the remaining larger angle.)

2 Doubling the cube.
Given a cube, to construct another of twice its volume. That is, in our algebraic notation: if s is the side of the given cube (with volume s^3), to construct a line, the cube on which would have the volume $2s^3$.

3 Squaring the circle.
Given a circle, to construct a square of the same area. Or, given a line which is the diameter of the given circle, to construct another line the square on which has the same area as the circle.

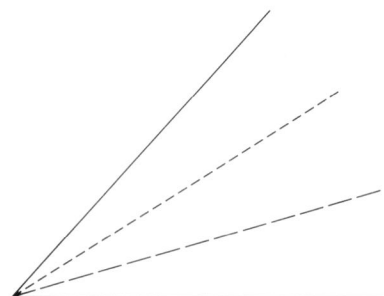

Figure 12   Trisecting the angle

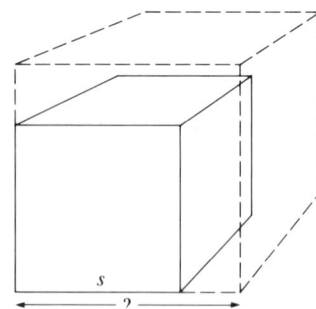

Figure 13   Doubling the cube

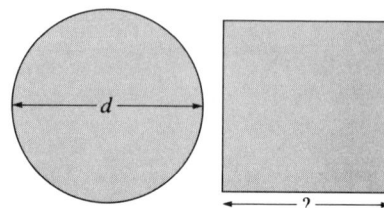

Figure 14   Squaring the circle

These three problems turned out to be enormously influential, leading to profound and fruitful mathematical studies in the course of attempts to solve them. It must have been somewhat frustrating initially for the fifth century mathematicians to find that these problems did not yield readily to the basic line-and-circle construction method. *Bisecting* an angle by straight-edge and compasses is very simple, as is trisecting a *line*, but a way of *trisecting* an *angle* by the same means could not be found. Doubling a *square* is straightforward—as you saw in the *Meno* extract, even a slave boy can do it—but doubling the *cube* seemed to be much harder. By the end of the fifth century, the problem of squaring the circle was sufficiently well known in Athens for the popular comic dramatist Aristophanes to make a joke out of it in his play *The Birds* of 414 BC.

How to bisect an angle is *Elements* II, 9 (**SB** 3.B3).

SB 2.C2

It was not until the nineteenth century that these problems were *proved* to be insoluble using line-and-circle constructions. The early Greek mathematicians, though, seem to have arrived rapidly at the practical judgement that they could not be done, and turned their attention to devising constructions that would produce solutions, albeit not by straight-edge and compasses alone. As a result of this activity, striking and important geometrical developments took place—notably, the early use of *conic sections*. So this enterprise marks a new kind of problem, one *created* by the methods of approach used within the developing research tradition. It was the *failure* of the basic line-and-circle construction method to solve these apparently elementary problems that led to much further enrichment of the mathematical methods, tools of analysis and proof methods of Greek geometry.

We now consider some of the issues arising from the ways in which solutions to the three classical problems were found.

Trisecting the angle
The earliest known method for trisecting an angle is through using a curve called the *trisectrix*, attributed to Hippias of Elis who flourished as a well known teacher of the Sophist school towards the end of the fifth century. The *trisectrix* is the line made by the intersection of two moving lines, in the following way. Starting from the square *ABCD*, imagine the line *BC* to move at a constant speed downwards (so that it ends up coinciding with *AD*) in just the same time as the line *AB* sweeps round a quarter of a circle, hinged as it were at *A*, so that *AB* also ends up coinciding with *AD* (Figure 15).

Plato in his dialogue *Protagoras* portrayed Hippias as a teacher in favour of compulsory education in the 'quadrivium' subjects; see **SB** 2.D2.

The *trisectrix* is the path traced by the point where these two moving lines cross. It makes possible the trisection of any angle, in the following way. Suppose the angle to be *EAD*, with *E* a point on the trisectrix. Then drop the perpendicular *EF* from *E* onto *AD*, and trisect it (trisecting a *line* can be done by straight-edge and compasses). So *G* is a third of the way along *FE* (Figure 16).

G then determines a point *H* on the trisectrix, level with it. Join up the line from *A* to *H*. Then angle *HAD* is one third of angle *EAD*. (It is not essential for you to know *why* this construction leads to the angle being trisected, but refer to the explanation in Box 2 if you are interested.)

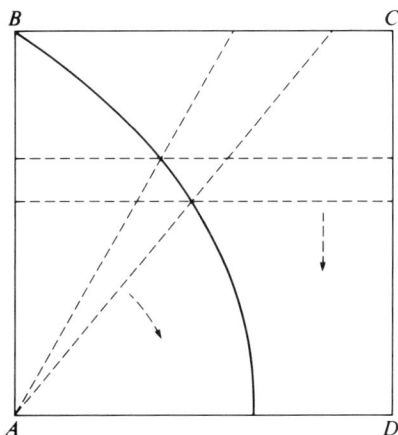

Figure 15

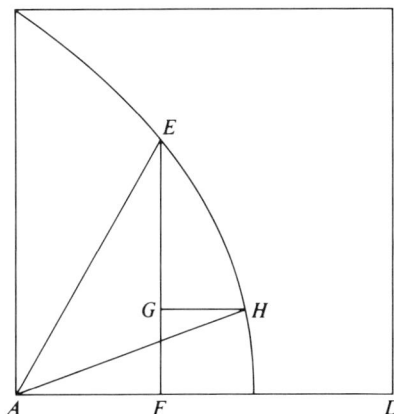

Figure 16

Our source for this construction of the trisectrix is the commentator Pappus (fourth
century AD), and the ascription to Hippias has been inferred by historians from
various other references by ancient writers. It is from Pappus, too, that we learn of a
neusis construction for trisecting the angle, which we look at now. (It is consistent
with our other knowledge and beliefs about early Greek mathematics to take this
construction as having been first devised in the fifth or fourth century BC,
somewhere in the period between Hippocrates and Euclid.)

This construction for the trisection of the angle will be presented here to exemplify
the *method of analysis*. In order to revise the notion of analysis, try this question.

Question 7 Suppose you were attempting to solve the problem of trisecting the
angle, so set out to analyse it (by the method of analysis). What would be the first
step in your analysis?

Pappus' text for this construction (in
SB 5.B4) contains only a synthesis.
We have reconstructed what the
preceding analysis may have looked
like.

Comment ⎯⎯⎯⎯⎯⎯⎯⎯⎯⎯⎯⎯⎯⎯⎯⎯⎯⎯⎯⎯⎯⎯⎯⎯⎯⎯

In Pappus' words, analysis is a method of 'taking that which is sought as though it
were admitted'. In this case you would suppose that some general angle *had* been
trisected, and carry on making deductions from that. ■

Let us carry out this process and see what happens.

1   Given any angle *ABC*, we suppose it has been trisected by the line *DB*.
(Figure 17)

This diagram is unilluminating as it stands, so we introduce a few more angles and
triangles to see what happens.

2   From *A*, drop a perpendicular down to *BC* and also draw a line from *A* parallel
to *BC*, meeting *BD* extended at *E*. (Figure 18)

(All of these constructions so far can be achieved with straight-edge and compasses.)

3   Finally join *A* to the mid-point of *DE*, at *G*. (Figure 19)

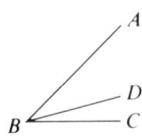
Figure 17

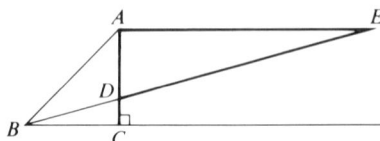
Figure 18

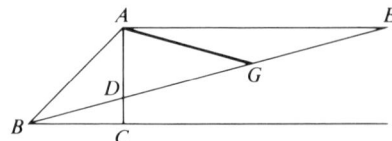
Figure 19

Now it is possible to make some inferences about the diagram, namely that

lengths *DG*, *GE* and *AG* are all equal.

(A similar result was seen in Hippocrates' first quadrature of lunes—by our
construction we have arranged that *G* is at the centre of the circle going through *D*,
A and *E*.) Further, we see that the *length of DE is twice that of AB*—this is shown in
Box 3.

So *if* the angle is trisected, *then DE* is twice AB. This is, then, the result of the analysis. With this as a starting point, the geometric argument would, we hope, work in reverse and we would end up by trisecting the angle. Such would be the process of synthesis. The condition we have reached is a *neusis* condition: *if* we can put a line twice the length of AB in such a way that its ends are on the two constructed lines *and* it is pointing ('*verging*') towards B, *then* we will have trisected the angle.

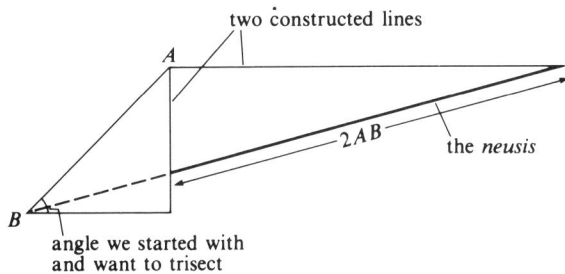

Figure 20

You can imagine how this could be done with a *marked* ruler: mark the length $2AB$ on it and move the ruler about until the condition is fulfilled. But this *neusis* (unlike the one in Hippocrates' quadrature) is impossible using an *un*marked straight-edge and compasses construction. So how can the trisection be logically justified? Pappus explained how the *neusis* can be constructed here, using the intersection of a circle and another curve called a hyperbola to determine the point E. But we shall not pursue this further at the moment as we return to the hyperbola and related curves later.

Doubling the cube

There is no record of why trisecting the angle originally arose as a problem, but with doubling the cube we have an embarrassment of colourful and informative stories. Two of these accounts are in the Source Book, from the commentators Theon of Smyrna (second century AD) and Eutocius (early sixth century AD), each claiming as *their* source works by Eratosthenes, the Librarian of Alexandria in the third century BC. The letter Eutocius attributed to Eratosthenes has been claimed as false (that is, Eratosthenes might not have written it), but it contains valuable information nonetheless. Please *read both of these texts now* (**SB** 2.F1 and **SB** 2.F3).

Question 8   What light do these accounts throw on the status (practical or theoretical) of the problem of doubling the cube?

Comment ———————————————————————————

The problem is presented as having originated in a practical way, through seeking to double an altar or a tomb. (You may have noticed that in the story about Minos, he is represented as making the same mistake as the slave boy in *Meno*, of seeking to double the sides in order to double the tomb.) The *motive* for this desired doubling is variously given. It is no longer clear to us why doubling an altar should be a good way of combating plague. The explanation ascribed to Plato is consistent with your study of *The Laws*, namely that he was ready to cite any unsolved

Unit 2, Question 5

mathematical problem to berate his fellow Greeks for ignorance and apathy towards mathematics. At all events, it seems that soon a whole battery of high-powered mathematicians—Archytas, Eudoxus, and Menaechmus, all friends or associates of Plato—were at work devising solutions to the related problem, due to Hippocrates, of 'two mean proportionals' (of which more below). So the problem was swept up into the highly theoretical geometric research tradition. It may in fact have originated there, as we do not know what credence to give to the tomb and altar stories. But the letter given by Eutocius goes on to point out the practical advantages of a mechanical construction to solve the problem, in terms useful for engineers or technologists.

Although we cannot infer from this that Greek engineers actually utilised mechanical solutions to the problem, nevertheless this is a useful reminder to us that a practical mathematical tradition did exist in Greek times. So although the logical status of the problem of doubling the cube is clear when seen within the theoretical geometric research tradition, practical knowledge or utility was later invoked too, in traditions where the high Platonic ideals held less sway. ■

You will see other examples later in the course of people making optimistic or implausible claims for the practical use of mathematics.

The most significant event in the history of doubling the cube was the advance made by Hippocrates of *reducing* the problem to that of finding *two mean proportionals*. We have seen examples of the *reduction* of a problem to other problems whose solution enables the original problem to be solved. For instance, we have seen how the problem of trisecting the angle can be reduced to that of finding a particular *neusis* construction. In the present connection Proclus commented:

> 'Reduction' is a transition from a problem or a theorem to another which, if known or constructed, will make the original proposition evident. For example, to solve the problem of doubling the cube geometers shifted their inquiry to another on which this depends, namely, the finding of two mean proportionals; and thenceforth they devoted their efforts to discovering how to find two means in continuous proportion between two given straight lines. They say the first to effect reduction of difficult constructions was Hippocrates of Chios, who also squared the lune and made many other discoveries in geometry, being a man of genius when it came to constructions, if there ever was one.

SB 2.F2

Before examining the connection between doubling the cube and mean proportionals, we should make an excursion into what proportionals mean. You may have noticed already that Greek mathematical language differed from ours in making much greater use of expressions of ratio and proportion. The first time you met this was in Hippocrates' quadrature of lunes, where he was said to have proved that 'similar segments of circles have the same ratios as the squares on their bases', by showing that 'the squares on the diameters have the same ratios as the circles'. So in the latter case, where we would write $d^2 \propto A$, or $A = \frac{1}{4}\pi d^2$ (for a circle of diameter d and area A), the Greeks had something like $d_1^2 : d_2^2 = A_1 : A_2$. Even though *in a sense* these all amount to the same thing, the differences in expression are rather important.

Furthermore, d^2 in the Greek case denotes the geometric square on the line d, not a number multiplied by itself.

In particular, this language enabled geometric entities to be treated as such, and their integrity preserved, rather than being swept up into the arithmetical language of numbers as we tend to do. Given two lines a and b, a *mean proportional* between them is a line x with the property that

$$a : x = x : b.$$

This is the equivalent, for lines, of what is still called *the geometric mean* between two numbers, namely the square root of their product. If we think of these lines as numbers then the property becomes

$$\frac{a}{x} = \frac{x}{b}, \quad \text{or } x^2 = ab, \quad \text{or } x = \sqrt{ab}.$$

From the $x^2 = ab$ formulation, we can see that geometrically the problem is that of constructing a square equal to a rectangle whose sides are a and b; the side of the square is then the mean proportional between a and b. This solution can be constructed with straight-edge and compasses, and appears as Euclid's *Elements* II, 14 (**SB** 3.C1): 'To construct a square equal to a given rectilinear figure'.

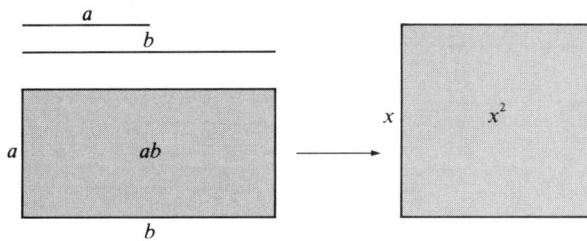

Figure 21

The problem of *two mean proportionals* cannot be solved by straight-edge and compasses, however. The problem is, given two lines *a* and *b*, to construct *two* lines *x* and *y* such that

$$a : x = x : y = y : b.$$

It was Hippocrates who seems to have established that if this problem can be solved, then the cube can be doubled. How this comes about may be readily seen: it is the case of two mean proportionals between the lines *s* and 2*s*, where *s* is the side of the cube to be doubled. Put into our algebraic language the condition is

$$\frac{s}{x} = \frac{x}{y} = \frac{y}{2s}, \quad \text{so} \quad \left(\frac{s}{x}\right)^3 = \frac{s}{x}\frac{x}{y}\frac{y}{2s} = \frac{1}{2}, \quad \text{or } x^3 = 2s^3.$$

Thus, if two mean proportionals can be constructed between *s* and 2*s*, then the cube on the first is twice the cube on *s*, which is what was wanted. In Greek mathematics thereafter, it appears that the original problem was always investigated in this reduced formulation.

Many different solutions, both mechanical and theoretical, to the problem of constructing two mean proportionals were put forward in Greek times. The sixth century commentator Eutocius described some dozen solutions. The theoretical constructions ranged from ones involving a *neusis* construction, to an astonishing construction of Archytas which determines the point of intersection of three surfaces in three-dimensional space. We look at just one solution, rather simpler than that of Archytas, and probably due to Menaechmus, a younger associate of Plato. This is a significant solution historically. From the work of Menaechmus in tackling the problem of two mean proportionals there may have arisen the Greek research tradition of investigating conic sections. We shall try to remain as faithful as possible to the Greek proportional language in describing what Menaechmus did, so you might like to consult Box 4 first, where his approach is outlined in modern algebraic language.

Box 4   A note on Menaechmus' solution to two mean proportionals

The original condition on two mean proportionals,

$$a : x = x : y = y : b, \quad \text{or } a/x = x/y = y/b.$$

can be rearranged to give an equivalent algebraic formulation as three equations:

$$x^2 = ay; \quad y^2 = bx; \quad xy = ab;$$

and from any two of these the solution (the values of *x* and *y*) can be extracted. Thus if we take the latter two equations we can turn the last into $y = ab/x$, put this value of *y* into the second equation and come up with $x^3 = a^2b$. So in the particular case $a = s$, $b = 2s$, this gives $x^3 = 2s^3$ which is what is wanted for doubling the cube.

Now what this means in geometrical terms is that the three equations determine curves, the intersection of any two of which will solve the problem. (The equations 'determine curves' in the sense that there are lots of values of *x* and *y* consistent with the condition $x^2 = ay$ (say), and drawing all these on a graph gives rise to a line which is the curve in question.)

17

What Menaechmus may have noticed was that each of these three sets of conditions refers to points on the surface of a *cone* which has been sectioned at right angles to a surface. For instance, if the cone is right-angled at the apex, then the curve of the cut face can be described by the relation $x^2 = ay$ (a *parabola*); whereas if the cone is obtuse-angled at the apex (greater than a right angle), the curve has a property describable as $xy = ab$ (a *hyperbola*).

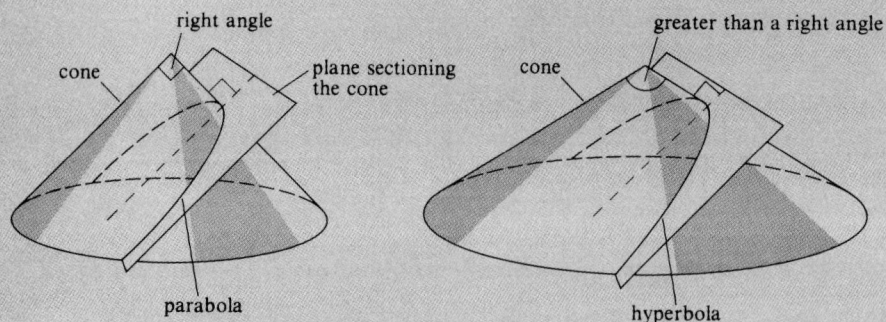

The upshot of this is that the problem of two mean proportionals can be solved by the intersection of a parabola and a hyperbola (or two parabolas), the intersection of the two curves corresponding to the values of x and y extracted from two of the equations above.

Menaechmus started his solution by supposing the problem solved (so, he used the method of *analysis*), and put down the two given lines a and b, and the two mean proportionals x and y, as on the diagram. Then from the condition

$$a : x = x : y = y : b$$

we can infer that the lines are related via

$$b \cdot x = y \cdot y,$$

so the point P Menaechmus recognised as lying on the conic section later called a *parabola*. The lines are also related via

$$a \cdot b = x \cdot y$$

which he recognised as relating to a *hyperbola*.

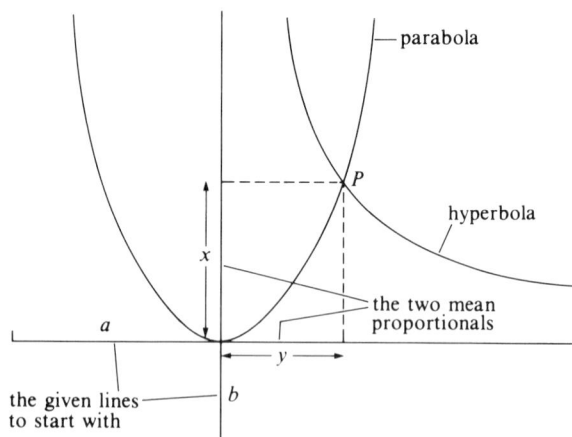

Figure 22

Thus the two mean proportionals are found to be where these two curves cut. The synthetic proof of this goes in the opposite direction: assume the parabola and hyperbola can be constructed (somehow), then reverse the argument. Thus the two mean proportionals can be found, and in the particular case $a = s$, $b = 2s$, the doubling of the cube has been achieved.

The analysis and synthesis of Menaechmus' solution, as given by Eutocius, forms **SB** 2.F4.

At some stage Greek geometers categorised the mass of problems they were learning to solve in various ways into three kinds.

Pappus is our source for this classification—see **SB** 5.B4.

Plane problems—those solvable by line and circle constructions (such as the problems of *Elements* I and II, discussed in the next section).

Solid problems—those solvable by conic sections (such as the doubling of the cube, and trisecting the angle).

Linear problems—those requiring curves other than conic sections. We have met one example of such a curve, the trisectrix.

The word *linear* is from 'line' in the sense of 'curve'. *Linear* has a different meaning in modern mathematics.

The third classical problem is a linear problem.

Squaring the Circle

This is an intrinsically more difficult problem than the other two, involving the relationship between areas bounded by curved and by straight lines, a relationship of paradox and perplexity in Greek (and later) times which we shall look at in the next unit. As you have seen, our earliest recorded result of Greek geometry was Hippocrates' quadrature of certain lunes—itself presumably part of an investigation of squaring the circle—which demonstrated that at least *some* things with curved sides are squarable. Ever since those early Greek times, 'squaring the circle' has been a by-word for the unattainable: the Italian poet Dante (1265–1321), for instance, wrote

> Like the geometer, who bends all his powers
> To measure the circle, and does not succeed.
> Thinking what principle he needs.

Dante, *Divine Comedy*

and in contrasting tone, an eighteenth-century English poet summarised this section succinctly in the lines

> Circles to square, and Cubes to double,
> Would give a Man excessive Trouble.

Matthew Prior, *Alma* (1717)

We deal briefly with just two aspects of the problem here: we show its reduction to another, and the use of the trisectrix to solve the problem.

Suppose, then, the circle has been squared. If some circle has radius R, then its square has area πR^2 (to use modern symbolism). But by a simple straight-edge and compasses construction any triangle can be squared. So this square is equal to a right-angled triangle whose sides are $2\pi R$ and R, the circumference and the radius of the circle.

Note that this analysis is one we have invented, using modern symbolism, for explanatory purposes. It does not derive from any known Greek text.

The area of any triangle is half the base times the height; in this case $\frac{1}{2} \times 2\pi R \times R = \pi R^2$.

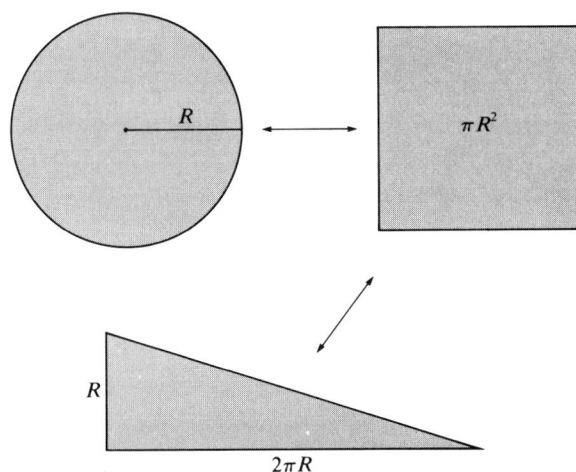

Figure 23

This modern analysis shows us that the circle can be squared if the circumference can be *rectified*—that is, provided one can construct a straight line equal in length to it. The result we have reached was proved by Archimedes during the third century BC and appears as Proposition 1 of his *Measurement of a Circle* (**SB** 4.A1).

Question 9   Look over this proof; read its statement and the first few lines to see Archimedes' approach. What logical form does the proof have? Is it an analytic or a synthetic proof?

Comment ──

The logical form of the proof is one which has not appeared much in the last few pages, a proof by *reductio ad absurdum*. He starts from the undeniable logical truth (which he does not write down, presumably as it is too obvious) that *either* the circle is equal to K (the triangle) *or* it is not equal to K, and proceeds to derive absurdities from the latter possibility, thus proving the former.

It is a synthetic proof, because it moves from its logical foundation by strict deduction towards the desired endpoint.   ■

So by the time of Archimedes, and probably rather earlier, the problem of squaring the circle had been *reduced* to that of rectifying its circumference. The latter can be done by using the trisectrix that served to trisect the angle. (Hence the more common alternative name for this curve, the *quadratrix*, so we shall change to this latter name when discussing the quadrature of the circle.)

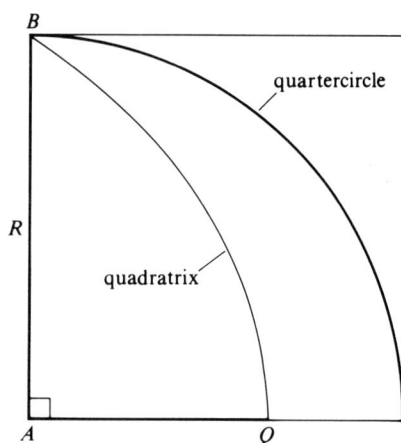

Figure 24

The way the quadratrix is used to rectify the circumference may have been devised by Hippias himself (the sources are a little ambiguous about this), or by Dinostratus, the brother of Menaechmus, in the middle of the fourth century (c. 350 BC). It consists of proving that

quarter circle : $AB = AB : AQ$.

So this gives the arc length of a quarter of the circumference in terms of the radius of the circle (AB) and the point Q at which the quadratrix cuts the horizontal line. We shall not go through the proof of this result which is also by *reductio ad absurdum*. If this proof is the original one, and was indeed due to Hippias, then it would be one of the earliest mathematical *reductio* proofs of which we have full record. Given the above result, the rectification itself (a straight line equal in length to the (quarter-) circumference) can be produced by straight-edge and compasses methods. For what the result says is that the circle's radius (AB) is the mean proportional between the arc length and the quadratrix distance AQ. Thus the problem becomes that of constructing a rectangle, one of whose sides is AQ, equal to the initial square on AB. Knowledge that this can be done is sufficient, so there is no need to go into details.

For this proof see **SB** 2.G4

But there is an aspect of this construction that it will be useful to finish on, as reminding us again of the extraordinarily high standards of logical rigour developed by the Greeks. Although it seems plausible, this construction of the quadratrix is invalid for rectifying the circumference—or so Pappus said, giving the objections of his friend Sporus. The reason for this is two-fold.

Question 10   Read the paragraph in which Pappus discusses these problems (**SB** 2.G4), beginning 'With this Sporus is rightly displeased . . .', and try to explain in your own words the logical difficulties involved.

His first point is that the end towards which the quadratrix is being devised (the quadrature of the circle) is assumed in its very construction. Remember that the quadratrix is defined by two simultaneous motions: the top line moves downwards in just the same time as AB pivots round on A so as to reach the bottom of the square together. So in order to construct this you need to know the ratio of their speeds, which is possible only if you already know the ratio of the side of the square to the quarter-arc length of the circle. But this is what you have set the construction up to find out! So the logic itself seems to be going round in circles.

The other objection is that the point labelled Q, where the quadratrix cuts the bottom line, cannot be identified—which is rather unfortunate as the length AQ is needed for the rectification. The problem here again arises from the original construction. Remember that the quadratrix is defined as the intersection of the two moving lines; but when they have both got to the bottom of the square they do not intersect in a unique point—they now lie on top of one another, and so in effect intersect everywhere (or nowhere, depending on how you regard intersection). If you look at the situation a moment earlier, it seems fairly clear where the quadratrix is heading (i.e. towards Q)—but that is not good enough by Greek proof standards. The location of Q would need to be justified by a further logical argument, which itself would have to rely on something equivalent to the rectification of the circle. ■

We shall see more of the kind of problem raised by the second objection in the next unit, when we look more fully at Greek approaches to the interesting difficulties and paradoxes of the behaviour of curved lines.

3.4 THE DEVELOPMENT OF AXIOMATISATION: EUCLID'S *ELEMENTS*

We have so frequently referred to results in Euclid's *Elements* that you should have a good idea of some of the things that appear. In this section we look at the work more globally, to understand the kind of enterprise it was. Euclid was not the first to write a work on *Elements*; we learn from Proclus that Hippocrates and others had done so previously, but their works have not survived. That tells us that Euclid was working within a well-established textbook tradition, apparently with such success as to overshadow all his predecessors.

Before looking at what Euclid's work contains, it will be useful to ask: what does the title *Elements* mean? Just as in modern English, where the word *element*—as in chemical element—can have a more specialised meaning than the adjective *elementary* (in the sense of simple), so too in classical Greece there was a range of meaning for the equivalent word *stoicheia*. Initially it meant any constituent of a line or row of things (so a soldier might be an *element* of a line of soldiers). Later, Menaechmus used it for any proposition used in the proof of another one (so in this sense 'the squares on the diameters have the same ratios as the circles' is an *element* of Hippocrates' quadrature of lunes). By the time of Euclid it meant, mathematically, a starting point for many other theorems. Proclus compares an *element* in this sense with a letter of the alphabet: from the letters of the alphabet (also called *stoicheia* in Greek) all words are constructed; and similarly the elements of (say) geometry are the foundational propositions from which the other geometric truths can be derived. Proclus went on to comment, in a passage which gives a strong indication of why Euclid was so valued by later mathematicians, that

This brief summary of changing meanings of the word *element* shows two things: that by such means we can pick up significant evidence about Greek development of proof through this kind of linguistic analysis; and that the name of Euclid's work is not an arbitrary label but a deep clue to the importance of its contents.

> It is a difficult task in any science to select and arrange properly the elements out of which all other matters are produced and into which they can be resolved . . . in general many ways of constructing elementary expositions have been individually invented. Such a treatise ought to be free of everything superfluous, for that is a hindrance to learning; the selections chosen must all be coherent and conducive to the end proposed, in order to be of the greatest

usefulness for knowledge; it must devote great attention both to clarity and to conciseness, for what lacks these qualities confuses our understanding; it ought to aim at the comprehension of its theorems in a general form, for dividing one's subject too minutely and teaching it by bits make knowledge of it difficult to attain. Judged by all these criteria, you will find Euclid's introduction superior to others.

SB 3.A1

Euclid's importance, then, lay in the choice and arrangement of 'the elements out of which all other matters are produced'. Throughout this unit we have seen examples of deductive proof, results shown to be true by logical inference from other results or assumptions. But Euclid's *Elements* is the first surviving example of a full-blown *axiomatic* system in mathematics, in which all the results are proved by reference back to foundational starting points. It is evident from the works of Aristotle, and, to a lesser extent, from Plato too, that such concerns were becoming more developed throughout the fourth century, at the end of which Euclid compiled what became the definitive *Elements*.

To understand the nature of an axiomatic structure better and to see what it was that Euclid achieved, both in logical terms and in its capacity to astonish and delight, let us approach it initially through the experience of Thomas Hobbes one day in about 1628. Please *read Aubrey's account* of that experience, **SB** 3.F2(a).

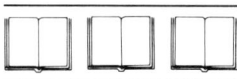

The proposition in question, *Elements* I, 47 is 'Pythagoras' Theorem', and to retrace Hobbes' steps will be a good way of investigating the structure of *Elements* I. *Read the demonstration* of *Elements* I, 47 (**SB** 3.B5), noticing that it refers you back to other propositions (by the numbers in square brackets) on which steps in the proof rely. These propositions in turn rely on other propositions, and one can construct a diagram summarising these chains of reliance, as in Figure 25.

We leave what Pythagoras' Theorem is *about* to the next section. Here we are concerned with what it reveals of the logical structure of a Euclidean Book.

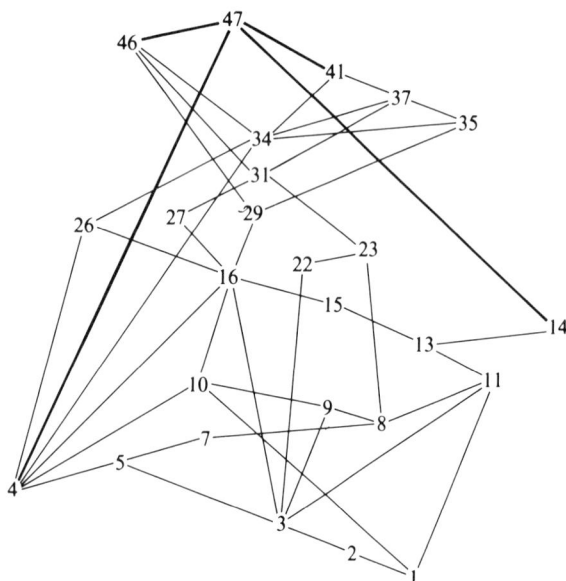

Figure 25

Propositions of *Elements* Book I which enter into the proof of Proposition 45. An upwards directed line indicates that the proof of the higher numbered proposition directly invokes the lower one.

From this it is clear that the logical structure is not a simple linear progression as implied in Aubrey's account, but a dense and complicated interaction of chains of reasoning. Aubrey conveyed the spirit of Hobbes' journey of discovery while somewhat oversimplifying it.

There are two things to notice from this exercise. One is that not all of the propositions in Book I have appeared on the diagram—in other words, they are not used in the proof of *Elements* I, 47. They are either of independent interest or not needed until a later Book. Indeed, Figure 25 is misleading if it is taken to imply that Book I is devoted to building up results leading to the proof of 'Pythagoras' Theorem'. There are stronger arguments (which we shall not go into) for viewing

Proposition 45 as the more significant mathematical result around which the Book was constructed. Proposition II, 45 is about transforming any rectilinear figure into another of equal area, and is central to constructions relating to quadratures.

Following on from our earlier discussions about the role of constructions, note in passing a significant difference between Propositions 45 and 47. The latter proves a mathematical theorem, as is indicated by the letters 'Q.E.D.' at the end. Proposition 45, by contrast, proves that something can be *done* ('To construct . . .'), as is indicated by 'Q.E.F.' at the end. Although constructions are involved in both proofs, that in 47 is the *means* of proof itself, whereas 45 is centrally devoted to justifying the validity of a construction. All Euclidean propositions fall into one or other of the 'Q.E.D.' or 'Q.E.F.' categories, which in some editions of the *Elements* are labelled 'theorem' and 'problem' respectively.

Q.E.D. stands for *Quod erat demonstrandum*, Latin for 'which was to be proved'; Q.E.F. stands for *Quod erat faciendum*, Latin for 'which was to be done'.

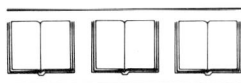

The second thing to notice from our Hobbesian exercise of working backwards through Book I is that the downward progression had to stop, of course, at propositions proved not from others but from the *postulates* and *common notions* at the start of the Book. Let us turn to these, for they are the heart of the axiomatic structure. *Read now Euclid's list of them* (**SB** 3.B1(a)).

It is not too important for our purposes what the difference between a *postulate* and a *common notion* is, but you may have noticed that the *postulates* have a geometrical subject matter, whereas the *common notions* are general truths or assumptions, applicable to wider inquiries too. We can think of them as geometrical and logical axioms, respectively. It is upon these, and the 23 *definitions* of Book I (together with further definitions introduced in later books), that the whole structure of Euclid's *Elements* is built.

Although much (if not all) of the content of the *Elements* seems to have been discovered earlier, it is generally agreed that Euclid himself either formulated the postulates, or selected them from all of those that may have been under discussion during the previous century or so. Euclid showed remarkable insight and intuition in making his choice of postulates, so that the entire superstructure was deducible from them and yet they were 'free of everything superfluous' (in Proclus' words). This is particularly noticeable in the case of Postulate 5, which is the *parallel postulate*. From virtually the moment of its formulation, mathematicians seem to have felt that it ought to be deducible from the other postulates, making it superfluous and not a postulate at all. Proclus declared roundly:

The parallel postulate *would* be superfluous if you could deduce it from the others.

> This ought to be struck from the postulates altogether. For it is a theorem . . . and requires for its demonstration a number of definitions as well as theorems.

SB 3.B1(b)

However, centuries of attempts to demonstrate it as a theorem were in vain, and eventually in the nineteenth century it was shown that Euclid had been right all along.

This is the subject of *Unit 13, Non-Euclidean Geometry*.

The previous postulates are interesting too, in that they define the scope of the *Elements*.

Question 11   Read the first three postulates, bearing in mind our discussion of construction techniques earlier in the unit. Would you expect to find in Euclid's *Elements* discussions of the three classical problems, or results about conic sections?

Comment ─────────────────────────────────────
Postulates 1–3 amount to setting line-and-circle constructions as the axiomatic foundation of the *Elements*. But there is no postulate allowing a *neusis* construction, say. So as the *Elements* consists only of results deducible from the axiomatic foundations given at the beginning, we should not expect anything other than what were called *plane* problems (see **SB** 5.B4). There are large and important areas of Greek geometry, such as the conic sections, which are not dealt with in Euclid's *Elements*.   ■

Euclid did write a separate work on conics, now lost.

It is remarkable, nevertheless, just how much does turn out to be encompassed by the framework of the *Elements*. Apart from our brief excursion into Euclidean number theory in the previous unit, we have been looking in detail only at Book I, and are far from exhausting the richness even of that! Books II–IV continue geometrical investigations in much the same vein as in Book I, with the geometry of circles beginning to occupy a more prominent role in Book III.

We discuss the content and interpretation of Book II in the next section.

Then in Book V the tone changes, building upon nineteen rather hefty definitions about magnitudes involving the notions of *ratio* and *proportion*. This *proportion theory* was probably due to Plato's pupil the great mathematician Eudoxus, and ensured a logical foundation for proving results about magnitudes (such as *lengths* and *areas*) in the abstract, whether they were commensurable or not. Book VI applies this to plane geometry, and Books VII–IX are the books on number theory that we looked at before. Book X is different again, the longest and most difficult of all; a working out of results on incommensurable magnitudes along the lines that seem to be hinted at in Plato's *Theaetetus* (see **SB** 2.E3). Finally, Books XI–XIII are mostly about solid (i.e. three-dimensional) geometry, the subject that Plato in *Republic* said had been neglected by mathematicians and should be investigated further.

The last few propositions of Book XIII, and thus in a sense the culmination of the whole work, is the construction of the five *regular solids*, and the demonstration that there are only five. (These are defined and described in Box 5.)

Box 5   A note on the regular or Platonic solids

A regular solid is one whose faces are all the same shape and size, and whose vertices all look the same. There are only five, namely:

Tetrahedron whose four faces are equilateral triangles;

Cube whose six faces are squares;

Octahedron whose eight faces are equilateral triangles;

Dodecahedron whose twelve faces are regular pentagons (five-sided figures);

Icosahedron whose twenty faces are equilateral triangles

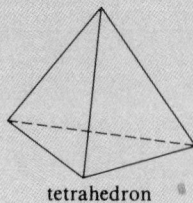

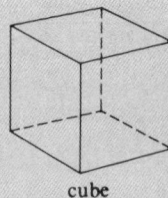

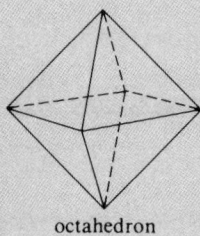

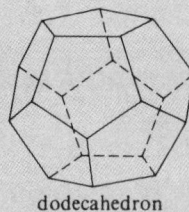

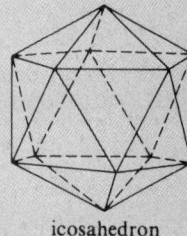

tetrahedron   cube

octahedron   dodecahedron   icosahedron

They are also called the *Platonic solids*, because of the important role they play in one of Plato's most influential dialogues, *Timaeus*. We shall see in a moment what that role is, but recall first that Proclus tells us that the whole purpose of the *Elements* was to lead up to their construction. In his historical summary he said

> Euclid belonged to the persuasion of Plato and was at home in this philosophy; and this is why he thought the goal of the *Elements* as a whole to be the construction of the so-called Platonic figures.

SB 2.A1

This tells us rather more about the importance of the regular solids for Neoplatonism than it does about Euclid or his *Elements*, however. For, bearing in mind that *Proclus* 'belonged to the persuasion of Plato and was at home in this philosophy', most historians have taken the view that in this remark Proclus was showing more enthusiasm than well-founded historical judgement. There are no good grounds for supposing other than that Euclid wrote the *Elements* for what seems its obvious purpose, to set down his view of the elements of the science. Indeed, it can be shown by examining the proofs of Book XIII that they could have been made considerably more succinct by using results from Book II. This rather suggests that Euclid cannot have had the Platonic solids firmly in view from the start, as he would have noticed this otherwise.

Book XIII of the *Elements* is generally agreed to be the work of Plato's friend Theaetetus, and it may be that he was the first to recognise that there are only five regular solids. It is worth noting that this result is of a kind we have not met before. It is one of the earliest examples of a 'classification theorem' in mathematics; that is, it is not about a single object, but a class of objects, and specifies completely all the objects with a particular property (in this case, all solids which are 'regular'). So the significant point here is not knowledge of any particular solid such as the cube, many of whose properties were doubtless studied earlier, but that a defining property could be singled out that was possessed by only the five pictured in the box. As William C. Waterhouse has put it,

> The real history of the regular solids therefore begins at the point when men realised there was such a subject. The discovery of this or that particular body was secondary; *the crucial discovery was the very concept of a regular solid.* Long familiarity has made the idea seem almost obvious, but it is not.

W. Waterhouse, 'The Discovery of the Regular Solids', *Archive for History of Exact Sciences* **9** (1972), p.214 (emphasis in original).

Theaetetus died in 369 BC, and only a few years later Plato incorporated a discussion of the regular solids into *Timaeus*, a work of great importance in the later developments of Neoplatonism (and thus, Western thought in general). In this text, Plato expressed most fully his conception of the creator-god as mathematician, who designed the world on mathematical principles. This explains, then, how the physical world of change, transience and decay is nonetheless rationally understandable; because it partakes, through the craftsman-like activity of its creator, in a pure mathematical design.

So *Timaeus* can be read as offering a theological view of how mathematics is applicable to the world; it is because the world is designed mathematically to begin with. (Of course, the problem this seeks to resolve could not have arisen in pre-Greek times, because until mathematics was taken to reside in some other-worldly realm of pure existence, the question of what it had to do with this world would have made no sense.) This perspective was to inspire many generations, right up to our own time.

Viewed in this light our earlier ambivalence between 'line-and-circle construction' and 'straight-edge and compasses construction' becomes easier to justify.

There are three extracts from *Timaeus* in the Source Book. The first describes the relationship between the four elements—earth, fire, air and water. It turns out that air and water are the two mean proportionals between fire and earth, so

SB 2.E5

fire : air = air : water = water : earth.

It is not necessarily appropriate to probe this formulation too closely—the significant point is that Plato explained the cohesion of the world through the language of proportion theory. The second extract is where the Platonic solids are introduced: he constructs them out of triangles and assigns one to each of the four elements, leaving the dodecahedron to represent the cosmos. The third extract is about the construction of the world-soul; it is a difficult passage that we shall not attempt to explain in detail, but it is well worth your reading through for the insight it gives on Platonic conceptions of the cosmos. In particular, the persistence of the *quadrivium* subjects as the most influential classification of mathematics in the educational tradition begins to make even more sense when you see the way Plato intertwines the arithmetic of musical harmony with astronomical geometry in a mysterious vision of cosmic mathematics.

Unit 2, Section 3

3.5 GREEK GEOMETRICAL ALGEBRA: *ELEMENTS* BOOK II

Historians of Greek mathematics have put much effort into trying to establish what can be learnt from Euclid's *Elements* about the mathematical developments of the preceding two centuries. It is, after all, both very substantial, and virtually our only source for many mathematical results and proofs reached during that period. By careful analysis of the text, and using all other available information, it has proved possible to come to various conclusions about which Books seem to have been developed at what time and by whom. It is generally accepted, for instance, that most of the *content* of Books I–IV must have been known to Hippocrates, though not necessarily the proofs found in Euclid and probably not its axiomatic form. Further, that some of these results are attributable to the Pythagorean school has been advocated by some historians, and vigorously denied by others. It would be too time-consuming, unfortunately, for us to enter into and help you to evaluate this and other historical controversies. But in this section we look at an important question on which historians are not unanimous: the question of what Euclid's *Elements* Book II is *about*.

The *Further Reading* at the end of *Unit 4* will give you some leads if you wish to follow these up.

First, we should see why this is a problem! On the face of it, a simple and by no means foolish response would be that Book II is about what it says it is about: propositions about rectilinear areas. *Glance through* the entries from this Book (**SB** 3.C1), and see whether this is your immediate impression too.

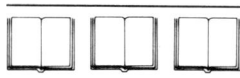

But on closer examination the question is more complicated. For consider Proposition 1. Apparently this amounts to showing that a rectangle is equal to the rectangles it can be cut up into by lines parallel to one side. Not only is this rather obvious, but the diagram alone would constitute a *diknume* proof—just looking at the diagram seems to make the result perfectly clear. The mystery deepens when we realise that this result is not referred to elsewhere in the *Elements*. So it does seem as though there is something here that warrants explanation of what Euclid intended to achieve through this proposition.

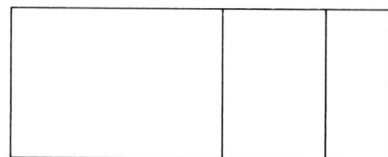

Figure 26

It has long been noticed (since at least the seventeenth century development of algebra) that when interpreted arithmetically or algebraically it is a rather less obvious result. Let us suppose, for example, that one side of the rectangle is 7 and the other 10, and we consider the latter as the sum of 1, 2, 3 and 4. Then the proposition says that

$$7 \times 10 = (7 \times 1) + (7 \times 2) + (7 \times 3) + (7 \times 4).$$

Or algebraically, let the lines have lengths a and b, and let b be divided into sublengths b_1 to b_4, so

$$b = b_1 + b_2 + b_3 + b_4.$$

Then the proposition can be interpreted as the 'distributive property of multiplication over addition', namely

$$a(b_1 + b_2 + b_3 + b_4) = ab_1 + ab_2 + ab_3 + ab_4.$$

A similar situation occurs with the other propositions in the book—Proposition 4, for example.

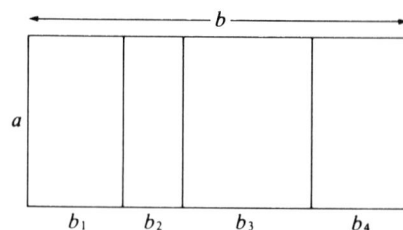

Figure 27

Question 12 Read the statement of Proposition 4 (**SB** 3.C1(c)) and try to formulate it in algebraic terms.

Comment ——————————————————————————————

If we suppose the initial straight line to be cut into lengths a and b, so that the whole is $a + b$, then the proposition turns into the familiar 'binomial identity'
$(a + b)^2 = a^2 + b^2 + 2ab$. ∎

However, you will be alert to the fact that wilfully changing a past mathematician's notation carries some danger of historical distortion. If Euclid had wanted to write down a piece of post-Renaissance algebra, then doubtless he would have done so. But the problem is deeper than symbolism alone; it is to do with style of thought and intention. To understand this better, let us become clearer about the Greek concept of *area*, the aspect of geometry that seems to be Euclid's overt subject-matter in Book II. When we studied the logical structure underpinning the proof of 'Pythagoras' Theorem' in Book I, one thing we did not look at was what the statement and proof were about; we repair that omission now.

Question 13 Here are three ways in which the result called Pythagoras' Theorem can be expressed.

(i) If the lengths of the sides of a right-angled triangle are a, b and c, then squaring each of these numbers it is always the case that $a^2 = b^2 + c^2$ (where a is the length of the hypotenuse).

(ii) If you construct a square on each side of a right-angled triangle, and add the areas of the squares on the two shorter sides together, that has the same value as the area of the square on the hypotenuse.

(iii) If you construct a square on each side of a right-angled triangle, then the squares on the two shorter sides can be cut up and rearranged so as to fit precisely onto the square of the hypotenuse.

Which of these does Euclid present as *Elements* I, 47? (You will need to study the proof a little to reach an answer—use the proof to interpret what the statement of the proposition means.)

Comment ──

I hope you could see that (i) is wrong: numbers have nothing to do with the matter. To distinguish between (ii) and (iii) is harder, so do not worry if you found it difficult to choose. In fact what Euclid does in the proof is best described by (iii). He cuts up the smaller squares into various pieces in a particular way and slides them about until they fit exactly onto the big square.

This is very important for what it tells us about the Greek concept of area. It is *not* to do with numbers (option (i)), nor to do with attaching numerical values to areas which can then be added up as numbers (option (ii)), but it is to do with area as two-dimensional stuff in its own right (option (iii)). ■

If, armed with this insight, we now go back to Book II, we can see that to some extent Proposition 1 illustrates the connection between lines and areas made in the opening definition of the book, of a rectangle being *contained* by two lines. If you look through the proof it is that definition which is repeatedly used. We could provide an algebraic translation of the proposition only by interpreting 'contained by two lines' (referring to a bit of two-dimensional stuff) as 'the product of two numbers' (which is another number). This is quite a dramatic restatement of Euclid's text, a changing of apparent thought style.

If it could be shown historically, however, that Euclid did have something of a more algebraic character in mind in Book II, then the content of the book would be easier for us to understand, even if the problem remained of why he wrote it down in so geometrical a way. With the greatly increased knowledge of Babylonian mathematics in the early part of this century, the answer to this conundrum seemed at hand. The Babylonian problems and solutions that were unearthed, translated and edited in the 1920s and 1930s were of a style that seemed akin to an arithmetised version of the concerns of Book II. This made possible a *historical* explanation of the earlier observation that Book II makes good sense when regarded as algebra. Thus in his influential history *Science Awakening* (1961), B. L. van der Waerden wrote of Book II

> We have here, so to speak, the start of an algebra textbook, dressed up in geometrical form . . .

> Presently we shall make clear that this geometric algebra is the continuation of Babylonian algebra. The Babylonians also used the terms 'rectangle' for xy and 'square' for x^2, but besides these and alternating with them, such arithmetic expressions as multiplication, root extraction, etc. occur as well. **SB** 3.G1

Let us look at another proposition in Book II from this perspective. Read Proposition 11 (**SB** 3.C1) to gain an initial impression of what it is about.

Here Euclid describes and performs a certain construction, but without revealing what it is used for or why it is of interest. One way of understanding it is to notice that later in the *Elements* Euclid returned to the idea (Book VI Proposition 30, **SB** 3.C4) and proved the construction again, this time using the more sophisticated techniques of Book V proportion theory. There it is called 'to cut a line *in extreme and mean ratio*', where the latter phrase means (Book VI Definition 2) 'as the whole line is to the greater segment, so is the greater to the less'. The construction is now known by a range of more fanciful names such as 'the golden section' and 'the divine proportion'. It arises in studying the regular pentagon as well as in other contexts.

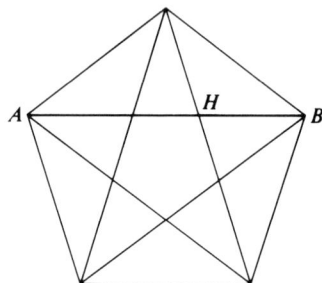

Figure 28 In the regular pentagon *H* divides the diameter *AB* in extreme and mean ratio

Question 14   Refer to the statement of *Elements* II, 11 and try to work out why this comes to the same thing as *Elements* VI, 30.

Comment ————————————————————

In order to cut a line in extreme and mean ratio, we want to find some point along it so that the two resulting segments have the property

   whole : greater = greater : less.

Thus, whole × less = greater × greater or, in the language of Book II, the rectangle contained by the whole and the less equals the square on the greater—which is *Elements* II, 11. ■

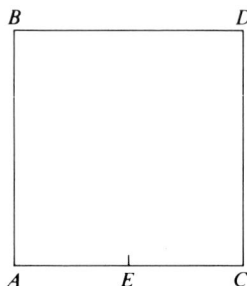

Figure 29

The proof of *Elements* II, 11 is a construction, and so falls into two parts as we have seen with earlier construction proofs: first doing the construction and then proving that this has the desired property.

We go through the construction, to get a feeling for the style of approach.

1   Given a line *AB*, construct the square on it (*ABCD*) and bisect the side *AC*. (Figure 30)

2   From the midpoint *E*, draw a circle with radius *EB*, and let *F* be the point where it cuts *CA* produced. (Figure 31)

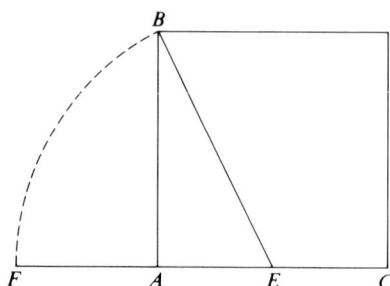

Figure 30

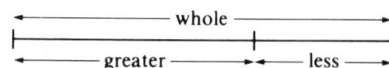

Figure 31

3 Draw the square on $AF(AFGH)$; the point H is then the one we want.
   (Figure 32)

That H is what we want is shown in the rest of the proof; the square $AFGH$ ('the square on the greater') equals the rectangle $HBDK$ ('the rectangle contained by the whole and the less'). (Figure 33)

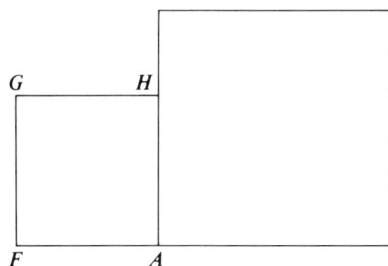

Figure 32

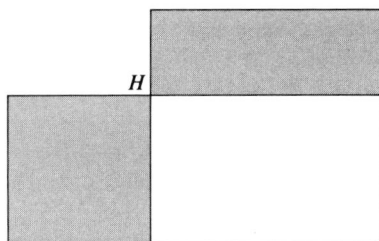

Figure 33

Now we have seen this through fairly Euclidean eyes, let us compare it with the 'geometric algebra' perspective. One textbook description of *Elements* II, 11 goes as follows:

> This is the first example of a geometrical solution to a quadratic equation. If we let $AB = 1$ and $AH = x$, the equation to be solved is
>
> $$x^2 = 1 - x \quad \text{or} \quad x^2 + x = 1.$$
>
> This problem is like the Babylonian ones discussed [earlier]. A Babylonian mathematician would have stated it in the form 'I have added up the area and the side of my square : 1', and would have solved it by taking half the coefficient of x, that is $\frac{1}{2}$. Euclid takes $AE = AB/2$. A Babylonian would then have formed the square of half the coefficient: $\frac{1}{4}$, then added it to the constant 1, obtaining $\frac{5}{4}$, and finally he would have taken the square root of this number (which he would not have been able to find exactly).
>
> Euclid adds up the squares, using Pythagoras's Theorem, and obtains EB^2, whose root is EB. The Babylonian would have subtracted $\frac{1}{2}$ from the square root. Euclid takes AF, which is the required solution. . . .
>
> Euclid's geometrical constructions follow exactly the same pattern as the Babylonian calculations, but Euclid always arrives at an answer whereas the Babylonian calculations only give an exact answer if the square root can be found exactly.

P. Dedron and J. Itard, *Mathematics and Mathematicians* (Open University Press, 1978) (original French edition 1959) vol. 2, p. 85.

Question 15 Does this seem a convincing account of *Elements* II, 11? Try to specify both the strengths and the weaknesses of this view.

Comment ————————————————————————————————
This account does have several advantages. The parallel drawn between Babylonian and Euclidean steps is neat and interesting; and the greater logical power of Euclid's method in reaching an *exact* (geometric) solution is well brought out (and would help explain why the Greeks formulated algebraic thoughts geometrically, if that is what they did).

On the other hand, the account is all built on the unexamined assumption that Euclid's proposition *is* 'a geometrical solution to a quadratic equation'. This is more problematic. For in the Greek text there are no equations, let alone quadratic ones, in sight. Euclid's proposition makes perfectly good sense in its own terms, as a geometrical solution to a geometrical problem, needing no appeal to external thought patterns to explain what is going on and why. ■

This debate took on an added liveliness in the 1970s, when a paper appeared in the respected journal *Archive for History of Exact Sciences* putting a view contrary to that of van der Waerden, Dedron and Itard. In forceful rhetoric, Sabetai Unguru expressed his belief that the 'geometric algebra' perspective was unhistorical, and was derived more from the algebraic training of modern historians than from anything in the Greek texts. To clarify and re-explain the processes which led to his adopting the 'geometric algebra' perspective, Professor van der Waerden replied:

An extract from Unguru's paper is **SB** 3.G2.

We . . . feel that the Greeks started with algebraic problems and translated them into geometric language. Unguru thinks that we argued like this: We found that the theorems of Euclid II can be translated into modern algebraic formalism, and that they are easier to understand if thus translated, and this we took as '*the* proof that this is what the ancient mathematicians had in mind'. Of course, this is nonsense. We are not so weak in logical thinking! The fact that a theorem can be translated into another notation does not prove a thing about what the author of the theorem had in mind.

No, our line of thought was quite different. We studied the wording of the theorems and tried to reconstruct the original ideas of the author. We found it *evident* that these theorems did not arise out of geometrical problems. We were not able to find any interesting geometrical problem that would give rise to theorems like II 1–4. On the other hand, we found that the explanation of these theorems as arising from algebra worked well. Therefore we adopted the latter explanation.

SB 3.G3

The issue has been widened by a further proposal, from the historian David H. Fowler. Note how van der Waerden said: 'We were not able to find any interesting geometrical problem that would give rise to theorems like II 1–4.' (An argument which Unguru felt to be more informative about Professor van der Waerden than about Euclid, see **SB** 3.G4). Fowler's claim is that such an 'interesting geometrical problem' does exist. Specifically, his work on the early Greek idea of *ratio* has suggested that early geometrical proofs here require the manipulation of figures found in Euclid's *Elements* Book II.

Further details are in Fowler's book *The Mathematics of Plato's Academy,* see *Further Reading, Unit 4.*

Let us draw the threads together by surveying what this episode of historical controversy has to tell us, about both Greek mathematics and the study of the history of mathematics.

First, it is clear that there is more to history than the theorems of the primary texts! Interpreting them, evaluating their purpose and significance, as well as situating them historically, can all prove to be difficult matters on which different views are possible.

Second, note however that the judgements of historians are not randomly imposed on the texts, but emerge through interaction with the evidence we have. Further, history being a rational enterprise, different views are argued for or against by means of the procedures of rational argument that are another aspect of the Greek legacy to the Western world.

Third, we can see the interconnectedness and complexity of judgements about the past, in that the implications of the debate on geometric algebra are much wider than simply the interpretation of Book II of Euclid's *Elements*. Unguru and van der Waerden seem, in fact, to have different views of the whole development of mathematics up to the seventeenth century.

Fourth, how might the debate about the status of Book II and 'geometric algebra' be resolved? If the claim that this is a rational argument is correct, then it should be possible to specify how to reach the best possible judgement on the matter. The debate would be materially advanced if there were more historical evidence about the transmission of mathematical ideas and practices from Mesopotamia to ancient Greece. Without some independent evidence that Euclid or his predecessors did know of the Babylonian work, the claim that they transformed it into geometrical garb must be viewed with some measure of caution or doubt.